Breeds of Hogs
Bulletin No. 124

by Louisiana Agricultural Experiment Station

with an introduction by Jackson Chambers

This work contains material that was originally published in 1910.

This publication is within the Public Domain.

*This edition is reprinted for educational purposes
and in accordance with all applicable Federal Laws.*

Introduction Copyright 2018 by Jackson Chambers

Self Reliance Books

Get more historic titles on animal and stock breeding, gardening and old fashioned skills by visiting us at:

http://selfreliancebooks.blogspot.com/

Introduction

I am pleased to present another title in the "Raising Pigs" series..

As with all reprinted books of this age that are intended to perfectly reproduce the original edition, considerable pains and effort had to be undertaken to correct fading and sometimes outright damage to existing proofs of this title. At times, this task is quite monumental, requiring an almost total "rebuilding" of some pages from digital proofs of multiple copies. Despite this, imperfections still sometimes exist in the final proof and may detract from the visual appearance of the text.

I hope you enjoy reading this book as much as I enjoyed re-publishing and making it available to fanciers again.

With Regards,

Jackson Chambers

INTRODUCTORY.

The purpose of this bulletin is to afford information to the farmers of the State concerning different breeds of hogs, and the crops on which they may be profitably raised.

Since the advent of the cotton boll weevil, and the barrier it has placed upon the successful growing of cotton as a single crop, our farmers have become more and more impressed, not only with the greater importance of diversification of crops, but also with the value of more and better live stock of different kinds. And, while inquiries have come to the Experiment Station for information regarding all varieties of farm animals, much more has been sought about hogs, and their food crops, than any of the others.

It is with the purpose, therefore, of giving a brief, but more or less accurate description of the different breeds of hogs, and the crops that our State is capable of raising to feed them on, and of endeavoring, in this way, to encourage the industry, to which our State is so admirably adapted, that the publication of this bulletin has been undertaken at the present time.

PART I. BREEDS OF HOGS.

By W. H. DALRYMPLE.

Without going back into the ancestry of the modern hog, it might be mentioned that, in Great Britain, where the stock from which the majority of our present breeds originated, it seems to have been the custom to divide their hogs into white and black breeds; and these, again, into large, middle, and small breeds.

Another division is into types, viz., the lard and the bacon types.

For the purposes of this bulletin, however, it may suffice to consider the hogs under color conditions, giving the particulars concerning each breed as we come to it.

BLACK HOGS.

THE POLAND-CHINA.

This breed is said to have originated in Southwestern Ohio.

DESCRIPTION AND CHARACTERISTICS.

Color—Black, with six white points, viz., feet, face, and tip of tail. The *face* is straight or very slightly dished, and the *ears* lop over about one-fourth to one-third from the tip.

A high quality specimen of the breed should be smooth throughout, with thick, broad, heavy *sides,* smooth *shoulders,* broad, heavy, plump *hams* extending well down to the hocks, with short legs and upright pasterns.

The *bone* is fine; the *tail* fine; the *head* and *ears* neat and attractive, and the *quality of the meat* is very good.

POLAND-CHINA.

The breed is characterized by *early maturing qualities,* and nearly perfect form of the *lard type* of hog, being blocky and compact.

The Poland-China seems to adapt itself to any conditions that furnish plenty of food for the production of quick growth; but also to a range of conditions comparable with other breeds.

The breed is admirable for crossing on common stock for grading purposes, and has been successfully used in different sections of Louisiana for a number of years.

Poland-China pigs, fed for market, may be made to weigh 200 pounds, or over, at six months of age. At one year old, males should weigh about 300 pounds, and sows, 250 to 275 pounds. While, in *breeding form,* at full maturity, males should readily weigh 500 pounds, and sows, 400 pounds.

The large type of Poland-China has more size; the flesh is coarser grained, and not as good in quality.

The *breeding qualities* of the Poland-China are considered fair.

THE BERKSHIRE.

This breed is of English origin, as the name would indicate, and is said to have been first brought to this country about 1830.

DESCRIPTION AND CHARACTERISTICS.

Color—Black, and, like the Poland-China, has six white points, viz., on feet, face, and tip of tail. The *face* is short to medium in length, and gracefully dished, and the *ears* are erect, or slightly inclined forward. The *back* is level and of moderate breadth, with considerable length of side. The *cheek,* or *jowl,* is full; the *shoulder* of medium thickness and breadth; the *ham* is deep and thick, extending well up to the body and down to the hock; the *leg* is medium to short, straight and strong, and widely set apart. The *bone* is of fair quality.

Although the Berkshire is classed as belonging to the fat or lard type, it is generally considered as a *medium* between the lard and the bacon types of swine.

The breed seems to be able to adapt itself to almost any environment, as it is found being successfully raised in the warmer sections of the South and Southwest, as well as in the more northerly parts of the country.

The *early-maturing qualities* of the Berkshire are classed as good; and as a *grazer,* it ranks high.

BERKSHIRE.

Pigs at six months old should readily weigh 175 pounds, and at one year old, about 300 pounds. In *breeding condition,* mature males should weigh about 500 pounds, and sows, 400 pounds.

For *grading purposes,* the value of the Berkshire male on common stock is undoubted. And where fair individuals of pure breeding are used, the Berkshire crosses well with the Poland-China, and other breeds.

The merits of this breed are already fairly well known throughout the State.

In general, the Berkshire is not lacking in *fecundity.*

THE ESSEX.

The native home of this breed is the county of Essex, England.

DESCRIPTION AND CHARACTERISTICS.

Color—Entirely black. The *head* is fine, rather short, with slightly dished *face;* the *ears* are thin and erect; the *jowl* is heavy and the *legs* are short and firmly boned. The *back* is broad and somewhat short, and the *sides* are deep and short. The *shoulders* are thickly fleshed, and the *hams* thick and deep.

The Essex is a small, compact, chunky hog on short legs. It is typical of the *lard type* of hog, except that it is small in size.

The breed is noteworthy for *early maturing qualities*—feeders maturing at six months of age.

As compared with the Poland-China and Berkshire, the Essex

ESSEX.

ranks among the smaller breeds. At maturity it is said to attain a weight of from 250 to 400 pounds, depending upon the care and treatment.

The *quality of the meat* is fine and well-flavored, with a tendency to fat over lean.

The *crosses and grades* may be of considerable comparative value. According to an English breeder, "There is probably no black pig which combines more good qualities as either porker or bacon hog than the produce of an improved Essex boar and an improved Berkshire sow."

The place for the Essex seems to be in the hands of the small breeder or feeder.

The Essex is not *perhaps so prolific* as some of the other breeds, but it cannot be looked upon as a shy breeder.

THE HAMPSHIRE OR THIN RIND HOG.

Originally this hog came from Hampshire, England, but in its present state of development, it may be said to be an American breed.

Description and Characteristics.

Color—Black, with a white band, four to twelve inches wide, encircling the body, and including the fore legs.

The *head* is small; the *ears* of medium length, inclining slightly forward. The *jowls*, or *cheeks*, are light; *back* of medium width. The *hams* are somewhat lacking in fullness compared with the larger American breeds; the *legs* are well set apart.

This breed is sometimes classed as a *medium* between the lard hog and the bacon hog, although it is generally considered as belonging to the former. It is, however, about medium in size.

A *cross* of the pure-bred male on common sows is thought likely to result in more prolific stock, leaning toward the bacon type.

The Hampshire is a good *grazer* and is *quite prolific*.

The *quality of the meat* is superior.

HAMPSHIRE OR THIN-RIND HOG.
(*Plumb's "Types and Breeds of Farm Animals."*)

RED OR SANDY HOGS.

THE TAMWORTH.

This is the extreme of the English bacon type of hog, and is one of the oldest and purest of British breeds.

DESCRIPTION AND CHARACTERISTICS.

Color—Red, or sandy, varying in shade from light to dark. Large in form; lean in type; and long in head, body and leg.

TAMWORTH.

The *snout* is long and tends to be straight, and the *face* is but slightly dished. The *ears* are large, and should be erect, or leaning slightly forward, and not breaking over. The *back* is narrow and long, and the *sides* long, and should be deep. The *shoulders* and *hams* represent the lean type.

The Tamworth is distinctly a *bacon type* of hog, and as such, ranks very high.

This breed is large in *size*, and can be made to weigh from 700 to 900 pounds, and even more. The average mature male will probably weigh about 600 pounds, and the sow, 450 pounds. At six months of age, pigs will weigh about 175 pounds.

The *early-maturing quality* of this breed is said to be inferior, but the *feeding quality* fairly good.

A favorite *cross* with some of the English feeders is between the Tamworth male and the Berkshire sow. One authority claims that Tamworth males bred on the fat type of American sows, should produce a very attractive, easy-feeding, highly marketable porker.

The Tamworth is *strikingly prolific*, and the males very prepotent. The breed has *good grazing* or *rustling qualities*.

THE DUROC-JERSEY.

The Duroc-Jersey is an American breed of swine, and in its latest improved state, is quite a modern hog. It is represented by some writers as originating in the Tamworth, the Red Berkshire, and the African or Guinea hog.

Description and Characteristics.

Color—Red or sandy, although the shades vary from light to dark. The *body* is rather short, with *legs* of medium length, and well-placed feet. The *head* is considered small in proportion to the size of the animal. The *face* is either straight, or

DUROC-JERSEY.
(*Courtesy of Prof. Dietrich, Ill.*)

slightly dished, and the *nose* is of medium length. The *ears* are of medium size, drooping forward, and the top third more or less breaking over. The *back* is of considerable width in contrast to length. The *body* often shows unusual depth. The *shoulders* and *hams* are rather heavy and thick fleshed.

In form of body, the modern Duroc-Jersey resembles the Poland-China more than the Berkshire.

This hog is classed among those of the *lard type*.

The *maturing qualities* of the breed are considered high class.

Pigs readily mature at six months old to dress out 175 pounds.

As to *feeding qualities*, the Duroc-Jersey classes with the Poland-China and Berkshire; and as a *grazer*, it seems suited to conditions favorable to other breeds.

The *quality of the meat* is regarded as good, compared with the Poland-China, when the same kind of food is used.

Pigs from common sows sired by Duroc-Jersey males, feed well, and when mated with Poland-China blood, are said to produce increased size of litters.

The *breeding* characteristics of this hog are considered distinctly superior.

The points of excellence in favor of the Duroc-Jersey are its *hardiness, growthiness,* and *prolificacy*. In the warm climate of the South, this hog has recently met with great favor, as it does not seem to be unfavorably affected by the dry, warm summers, especially as regards skin troubles, which the writer is inclined to believe is somewhat due to its color.

WHITE HOGS.

THE YORKSHIRE.

This is an old English white breed of hogs. Its blood is said to have entered into the formation of practically all breeds, either of English or American origin.

There are three different varieties of the Yorkshire, viz., the Large, Middle or Medium, and Small. The Large Yorkshire is a *bacon hog*, while the Small Yorkshire is of the *fat or lard type,* and the Middle Yorkshire is a *medium* between the two.

THE LARGE YORKSHIRE.

Description and Characteristics.

Color—White. The *head* is medium in length, with but little upturned curve; the underjaw is broad and strong; the *ears* incline to be heavy and droop forward. They should be fine, of medium size and be carried well upward and only slightly pointing forward.

The *body* should have considerable length; the *back* should be of fair and uniform width with considerable depth, the *sides* being long and deep at the flanks, and full between shoulder and hip.

LARGE YORKSHIRE.
(*Courtesy of Prof. Dietrich, Ill.*)

The *hams*, although not extremely fat and heavy, should be of good size and thickness, with the thighs well carried down.

In *size*, the Large Yorkshire stands in the first rank. Sows of this breed have been known to weigh considerably over 1,000 pounds.

The *early maturing qualities* of this breed are not pronounced.

Cross-breeds of this breed are of very superior type, especially from our typical American sows. As a *bacon producer*, this hog ranks high; and as a *breeder*, it takes a first place. Its *feeding qualities* do not seem to have given perfect satisfaction in this country; neither does the pure-bred *graze* equally well with our more common breeds.

THE MIDDLE YORKSHIRE.

Color—White. This hog is said to be the outcome of the cross between the Large and the Small Yorkshire. It is of smaller and *fatter type* than the Large White. Its *face* is more dished, its *back* is broader, and it more nearly approaches American ideals of form. The Middle Yorkshire is not a recognized breed in America, although it is in Great Britain.

MIDDLE YORKSHIRE.

THE SMALL YORKSHIRE.

The origin of this, the smallest of the Yorkshires, is, in England, obscure. It is assumed to have come from Chinese stock, although different today from the early Chinese type.

DESCRIPTION AND CHARACTERISTICS.

Color—White, except for occasional black spots.

The *face* is very short and broad, and often dished to such an extent as to point the end of the nose upward. The lower jaw, also, is much curved upward; the *ears* are short, fine and erect, and pointing forward; the *jowl* is very round and highly developed.

SMALL YORKSHIRE.
(Plumb's "Types and Breeds of Farm Animals.")

The *back* is very broad, short, and thickly laid with flesh or fat.

The *hams* and *shoulders* are heavy and full, and the *bone*, *hair*, and *quality* are refined.

The breed ranks as the smallest in this country, but it has great breadth and depth for its size.

The matured pig will weigh about from 180 to 200 pounds, although they have been made to weigh nearly 300 pounds at 15 months of age.

The *early maturing qualities* of this breed rank very high; they *fatten* readily, and are said to be good *grazers*.

The *quality of the meat* is good, but it contains a high percentage of fat.

To obtain good results from *crossing*, the Small Yorkshire should be mated to animals of the larger breeds.

As *breeders*, they rank as medium, with a tendency to small litters.

THE CHESTER WHITE.

The original stock of this breed was imported from England, but the breed itself originated in Chester County, Pennsylvania.

Description and Characteristics.

Color—White, with occasional black, or bluish-black spots on the skin.

The *size* ranges from a large hog to one of medium size.

The Chester White belongs strictly to the *fat or lard type*, producing a relatively large proportion of fat meat to lean in the carcass.

The *face* is straight, with the *nose* tending to be long and narrow, and the *ears* are drooped forward, breaking over from one-half to one-third of their length.

CHESTER WHITE.
(*Courtesy of Prof. Dietrich, Ill.*)

The *body* is blocky in form, although not very long. with heavy *hams* and smooth *shoulders;* broad *back*, and smooth throughout.

Mature males in fair flesh will weigh about 600 pounds, and sows about 450 pounds; while barrows at 6 to 8 months old should feed to weigh 350 pounds.

As a *feeder,* the Chester White ranks high; and is considered one of the most *prolific* of the heavier-fleshed types.

As a *grazer,* this hog will do well on good pasture, and compares favorably with other breeds.

Cross-bred and grade Chester Whites make excellent feeders.

The *quality of the meat* ranks as good.

On account of their color, they, like other white-skinned breeds, are liable to sun-scald, and other skin troubles, from the hot sun in our Southern climate.

THE CHESHIRE.

The native home of the Cheshire is Jefferson County, New York State. It is the outcome of crosses between the Large Improved Yorkshire and Suffolk breeds upon native white hogs.

DESCRIPTION AND CHARACTERISTICS.

Color—White. The *face* is somewhat dished, although not extremely so; the *ears* are small and fine, stand erect or point slightly forward.

The *back* should be long, broad and straight nearly to the

CHESHIRE.
(*Courtesy of Prof. Dietrich, Ill.*)

root of the tail. The *body* as a whole has considerable length, but frequently lacks depth.

The *hams* and *shoulders* are regarded as thick-fleshed and well-developed; the *leg bones* show considerable refinement.

The *size* of the Cheshire is about medium, and as a breed, closely resembles the Middle Yorkshire in form.

The *quality of the meat* ranks high.

Cross-bred and grade Cheshires are not very common, but are looked upon with favor in some localities.

As a *feeder and grazer*, the qualities of this breed are not as yet very well known.

The *breeding qualities* of the Cheshire are above medium.

The popularity of this breed is largely local, it being one of the least known breeds in the country.

THE VICTORIA.

The origin of this breed of hogs is accredited to two sources, which has resulted in the Davis Victoria and the Curtis Victoria, the former an Indiana product, and the latter produced in New York State.

VICTORIA.
(*Courtesy of Prof. Plumb, Ohio.*)

Description and Characteristics.

Color—White. The *head* is moderately broad; the *face* medium dished; the *ear* small to medium in size, and carried erect.

The *body* is broad and deep; the *back* level, and the *tail* set on at a line nearly level with the back.

The *hams* and *shoulders* are considerably thick and full, and the length and depth of *side meat* is very good. The length of *leg* is moderate, and the quality of *bone* is fair. In form, this hog somewhat resembles the Middle White Yorkshire.

In *size,* the Cheshire ranks as medium with the Poland-China and Berkshire. At maturity, the males should weigh about 600 pounds, and the sows 450 pounds.

As a *feeder,* this breed has not been extensively tried, as its distribution has been limited chiefly to Indiana, Ohio, and Illinois, although herds are found in some other Mississippi Valley states.

The *quality of the meat* will rank well among that of other breeds.

The *cross-bred or grade* Victoria has not been extensively tried, and its value is not commonly known.

The *breeding qualities* of this hog are considered very good.

THE COMMON SOUTHERN HOG.

THE "RAZORBACK."

The hog which we in the South know as the "Razorback" is a representative of the unimproved breed in this country, and is confined, almost entirely, to the more southern states.

Description and Characteristics.

Color—Varied. As a breed the Razorback is a *long-bodied, long-legged, thin, long-nosed* hog, exceedingly *hardy* and with remarkably *good foraging powers,* but *too slow in coming to maturity,* and *too light in weight* to compare with the modern and improved breeds of swine. In describing this breed of hogs, a writer adds: "It has no place in modern agriculture."

If we consider the Razorback from the standpoint of a breed, only, we fully agree with the writer just quoted. But this hog is susceptible to much improvement through an intelligent

THE "RAZORBACK."

system of grading by the use of males of improved breeding, and especially of the larger kinds, on the best of the native females. In this State, the "woods are full" of razorbacks, and, although they "have no place in modern agriculture," our problem is how to get rid of them, and at the same time make the most out of them. In suggesting methods of improvement in all varieties of our native live stock, the writer has always recommended the grading-up process by the use of pure-bred males; and in the case of our hogs, we have the same advice to offer. The first step should be the conversion of all native male pigs into barrows. The second should be the selection of the best of the young native sows for breeding purposes. And the third should be the purchase and use of pure-bred males of whichever breed and type the owner prefers. In a few generations of judicious selection and grading, we would be able to bring our native hogs up to a class that would be more profitable, and that would find a ready market at a much younger age than at present obtains. Except in the case of those who desire to go into the business of breeding and raising pure-bred hogs for sale, it is unreasonable to expect that the ordinary raiser of swine can afford to purchase all pure-bred animals.

It is too expensive a proposition. But, in his native sows, he has already got the basis for improvement; and if he will only carefully select these, and use pure-bred males of the improved types, it will only be a matter of a few generations of grading-up before he has a class of hogs that the best markets will be very glad to secure. This seems the most intelligent and rational way of "getting rid" of our Razorbacks, which the writer quoted says, "have no place in modern agriculture."

AS TO THE "BEST" BREED OF HOGS FOR THE FARMER.

The question is often asked, "What is the best breed of hogs?" In answering this question, it may be said there is no one breed of hogs that is "best" for every farmer. That which is best for one may not be best for another. In a general way, the "best" hog is the one the farmer likes, provided it is what his market demands. Should his market want a hog of the bacon type, then one or other of the breeds of that type would be best in his case. On the other hand, should the most marketable be a fat, or lard, hog, then the most profitable animal for him to raise would be one or other of the lard types.

This is a matter which the farmer will have to decide for himself.

After the farmer has decided upon the breed of hogs which he believes to be the best and most profitable for him to raise, however, he should then stick to that breed and endeavor to develop it to its most perfect condition.

There has been a tendency, up to the present, to change breeds frequently, and upon the slightest suggestion, whether with apparent reason or otherwise, before adequately determining the suitability of each as to the needs of the farmer, or the requirements of his market. This is to be deprecated, as it is not possible to get the most out of a breed by continually changing from one to another.

The most famous individuals, and the most famous herds of animals, have been built up, developed and perfected, only by sticking to the breed and getting the most out of it.

There is plenty of room, and sale, for all the different breeds of hogs; but it is useless for one individual to try to raise, and bring to their most perfect development, all of them.

In the opinion of the writer, there will be an increasing demand in the State for pure-bred hogs, of all kinds, for breeding and grading purposes; and those who make a specialty, each of his own particular breed, will reap the benefit of increased prices for animals of the highest excellence.

STATE REGULATION CONCERNING THE IMPORTATION OF HOGS.

In order to protect the hog industry in the State by preventing diseases being introduced through imported animals, the Louisiana State Live Stock Sanitary Board has a regulation bearing upon this subject, which we here quote as a matter of information, and to prevent possible inconvenience, to prospective purchasers of hogs in states outside of the State of Louisiana:

Regulation 5 (of the Louisiana State Live Stock Sanitary Board)—"All live stock, when brought into the State of Louisiana by any person, persons, firm, corporation, railroad, or other transportation company, shall be accompanied by a certificate of health, and said certificate of health shall state that said animal or animals is, or are, free from infectious, contagious, or communicable disease. Said certificate must be made by a qualified veterinarian immediately after he has personally examined the live stock, and within twenty-four hours before the live stock have been shipped into the State of Louisiana. Said certificate shall be attached to and accompany the shipping bill of the live stock to the place to which the live stock are shipped, and the owner of the live stock, or the agent of the transportation company, shall mail or send such certificate to the Secretary and Executive Officer of the Louisiana State Live Stock Sanitary Board, at Baton Rouge, immediately following the arrival of the live stock at its place of destination; provided, that the provisions of this regulation are not intended to apply to live stock which are to be used for immediate slaughter, but which must be free from all infectious, contagious, or communicable diseases."

Notwithstanding the above regulation to prevent the importation of diseased animals into the State, it is earnestly recommended, as a further precaution against hog-cholera, that all hogs purchased in other States, or, in fact, anywhere outside of the purchaser's own premises, be strictly quarantined, after their arrival, for at least two weeks before being placed with the herd, and that they be sprinkled, twice a week, or so, with a 2 or 3 per cent solution of any of the standard coal-tar dips, and the quarters kept thoroughly disinfected.

The purpose of the above recommendation is to allow time for the cholera to develop while the animal, or animals, is in quarantine, should the disease be present in latent form, and thus prevent its transmission to the herd, which in many cases might readily occur in the absence of the precaution mentioned.

It might be stated, also, that the above mentioned solution will be found useful for sprinkling or dipping pigs infested with lice.

In compiling the information with regard to the different breeds of hogs, the following authorities have been consulted and quoted:

"Types and Breeds of Farm Animals" (Plumb).
"Farmer's Cyclopedia of Live Stock" (Wilcox & Smith).
"The Study of Breeds" (Shaw).
"Swine" (Dietrich).
"Live Stock of the Farm" (Pringle).
"The Live Stock of the Farm" (Edited by J. Chalmers Morton).

A USEFUL MIXTURE FOR HOGS.

Take

    6 bushels of corn-cob charcoal, or,

    3 bushels of common charcoal,

    8 pounds of common salt,

    2 quarts of air-slacked lime, and

    1 bushel of wood ashes.

Break the charcoal well down, with shovel or other implement, and thoroughly mix all of the ingredients together. Then take 1¼ pounds of copperas (sulphate of iron), dissolve it in hot water, and with an ordinary sprinkling can, sprinkle the solution over the entire mass, and again thoroughly mix. Put the mixture in self-feeding boxes, or boxes protected from the weather, and place them where hogs of all ages may eat of their contents at pleasure.

Quantities, either greater or smaller than that given, may be prepared by simply observing the proportions of the different ingredients.

HOG CHOLERA SERUM.

All inquiries regarding hog cholera serum should be addressed to the Secretary, Louisana State Live Stock Sanitary Board, State Capitol, Baton Rouge.

PART II. THE BEST CROPS TO GROW FOR HOGS.

By W. R. Dodson.

In those States where corn is the chief crop of the farm, hogs are raised largely on this grain. While excellent corn can be grown in every portion of Louisiana, hogs can be grown most profitably under a system that permits the hogs to harvest a large portion of their feed in the form of green forage. While grain fed hogs grow more rapidly than those that are supplied with abundance of forage and a small quantity of grain, the latter system of feeding gives a more profitable hog in Louisiana. It is the purpose of the following discussion to give what the writer thinks has been demonstrated to be the best and most economical methods of growing and feeding hog crops.

Hogs should be bred so as to give one litter of pigs in the early fall, and the second in late spring. Good management will accomplish this distribution of breeding. The fall litter should be carried through the winter and spring largely on green crops that may be grazed, and brought to maturity during the late summer and fall at a year old or slightly more, on matured crops which the hog is allowed to gather. The spring litter should be maintained largely on green crops until about the first of August, and marketed at six to eight months old, with more grain and concentrated feed than received by the first litter. Exclusive feeding for a period of two weeks on concentrated dry feed just prior to killing, will give meat equal to that fed exclusively on grain throughout the fattening period.

Of the crops that can be grown to furnish good grazing during the winter months, rust proof oats, southern grown rye, barley, dwarf Essex rape, the clovers and alfalfa are all available in some localities, and most of them are available in all localities in Louisiana. For mature forage for the fattening period, cow peas, soy beans, velvet beans and peanuts are easily the leaders. Of the root crops that may be grown, sweet potatoes, artichokes and cassava are available, and stock beets, rutabagas and turnips may be used from early winter to midsummer as supplemental feed. Of the concentrated feeds available for the finishing period, and for supplemental forage when it may be desirable, we may use our corn, oats, rice polish and rice bran, and a very small quantity of cotton seed meal. No doubt peanut meal may become available in the near future, if the crop of peanuts increases as the interest in this crop at present indicates it will.

OATS FOR WINTER GRAZING FOR HOGS.

The genuine rust proof oats, from Louisiana grown seed, may be sown as early as the second week in September, if moisture of soil is sufficient to germinate the seed, or they may be sown as late as the latter part of November. From early sowing good pasturage will be furnished in four weeks from date of sowing. We prefer to sow not less than a bushel and a half and not more than two bushels per acre. The land should be well prepared and the seed covered to a depth of one and a half to two inches. Late sowing should not be covered so deep. If a grain drill is not available, a good disc harrow serves the purpose well to cover the seed. An ordinary iron tooth harrow will serve as a last resort. Sowing seed in advance of the turn plow is sometimes practiced, but the results are uncertain, and this is not considered good practice.

As to whether or not land should be pastured during wet

weather, depends upon the quality of the land. If the soil is deficient in vegetable matter, and has a large amount of clay, it is liable to become very hard in dry weather, if pastured during wet weather. Little if any Louisiana land will be injured by pasturage when the soil is dry enough to permit of plowing. If land is to be selected especially for winter pasture crops that do not form a permanent sod, the sandy soil should be selected.

If fall sown oats are not pastured later than the first of February, or even the middle of February, a good crop of grain will be produced in May, after which other crops may be planted.

We are frequently asked for information as to how many hogs an acre of oats will maintain during the winter. This question can be answered only with very wide limitations. It depends upon when the oats are sown, how rich the land is, how cold the winter is and how much wet weather we have. Oats sown in early October on good land, in an average year, should maintain four or five sows with their fall litters of pigs, if the mothers are fed a moderate amount of concentrated feed at first, diminishing the amount as the pigs eat more oats. One-third to one-half a normal feed of grain to the mothers will give a good flow of milk when the oat grazing is good. When the pigs begin to eat, a very small quantity of rice polish or bran will add much to their thrift.

It is the opinion of the writer that it is better to take the hogs from the oats and allow a crop of seed to mature, than to allow the hogs to continue to graze the crop until summer. This matter will be referred to again under the discussion of clovers.

The question is frequently asked, is not the oat crop an exhaustive one on the soil? It makes only a moderate draft on soil fertility. If the oat crop is followed by cow peas or peanuts, the land will not be depleted by the production of the two crops more than would be represented by the removal of the phosphate and potash. No just objection can be made to oats on this point, as the cow pea and the peanut greatly enrich the soil in nitrogen content.

As a grain crop, oats should produce twenty to thirty bushels per acre in the hill lands, from thirty to sixty bushels in the bluff soils, and greater yields in the alluvial lands, especially in the northern portion of the State.

In the purchase of seed oats, Louisiana grown seed should be secured by all means. Texas grown seed are very apt to contain Johnson grass seed, and frequently oats sold as Texas Rust Proof Turf oats, are only partly resistant to rust.

Spring-sown oats do not ordinarily give satisfactory results for any purpose.

If fertilizer is to be used, a moderate quantity of nitrogen and phosphoric acid is desirable. One hundred and fifty pounds of high grade acid phosphate and the same amount of cotton seed meal per acre, on the poor quality of land should start the crop off well. In the sandy lands of the hill parishes, a light top dressing of nitrate of soda, in the spring, will give good results in many cases.

After extended experimentation and observation, the writer is convinced that there is no better winter grazing crop for hogs, that is available to all the State, than rust proof oats, sown in the early fall.

CLOVERS.

Clovers should be sown in the early fall, if the season is favorable. They are more likely to be killed by drought than are the crops that grow more vigorously. Early October is the best time, if the season is favorable. The Medium Red is the best variety for the bluff lands and the alluvial soils, and the better quality of clay soils of the hills. The Crimson Clover is the best for the sandy soils of the hills, and the Alsike will grow in some stiff, poorly-drained lands not suited to the other clovers. It is advisable to sow about twelve pounds per acre. At the present time the seed are unusually high in price, costing from seventeen to twenty cents per pound. Clovers may be sown alone or with oats. If sown with oats use about eight pounds per acre. When desired for hog pasture to be used after the animals are taken from oat pasture, we prefer clovers sown alone. The land should be well prepared, the seed sown broadcast and harrowed in, so as to not cover the seeds more than an inch deep. If sown with oats it is best to sow the oats with a drill first and then sow the clover and harrow lightly. If oats are not sown with a drill, clover and oats may be sown and harrowed in together, but we prefer separate sowing and harrowing. Clovers will furnish good grazing land by the time the hogs are taken from the oats. They will supply the very best quality of grazing until the native grasses and white clover in pastures of sod are good, and then make a good crop of hay in late May. If clover is sown in early fall and not pastured it will be ready for harvest in early May, when the season is likely to not be very favorable for hay making. Pasturing retards the development and brings the hay cutting time at a more favorable season. For this reason clover for pasture for hogs fits well with oats, as well as furnishing an abundance of most excellent feed. If permanent sod of Bermuda grass and white clover is available, and it should be provided, the hogs can be gradually cut off the cultivated clover pasture in time to allow a good hay crop to mature. If the clover is not grazed too severely, and the hogs are removed in the early part of April, a maximum of about two tons of clover hay may be harvested about the last of May or the early part of June, and a small second cutting secured in about five to six weeks later. The clover will not survive through the summer, and it is then best to plow the land and plant a summer crop. Clover hay is a very fine hay for dairy or beef cattle, can be fed to advantage to hogs, but when fed to horses produces an excessive flow of saliva, which becomes disagreeable to parties handling the animals. No other ill effects follow. Clover hay is more difficult to cure than our Lespedeza, or Bermuda grass.

All clovers greatly enrich the soil, and should be much more extensively cultivated for grazing purposes, especially for hogs.

When oats and clover are sown together and harvested at the time the oats are in the "dough" the quality of hay is very superior, and is much more easily cured than is clover alone. If this hay can be chopped in a machine before it is fed, there is little waste and the combination is most excellent for all farm animals.

DWARF ESSEX RAPE.

Dwarf Essex Rape furnishes good winter grazing for hogs, on rich land. It is useless to plant it on poor soils. The seed should be sown broadcast, at the rate of five to ten pounds per acre, and harrowed in. Early October is the best time to plant, if the season is favorable. This crop makes a luxuriant growth throughout the winter, but does not give the returns in feed value that its appearance leads one to expect. Oats have given us more money returns, and oats are preferred by hogs as a grazing crop There is no return from a rape crop after the grazing, as there is with oats. In short, Dwarf Essex Rape will give a good crop on rich lands, if sown early, but oats will meet every requirement that rape will, and will generally give better returns.

RYE.

Rye affords good winter grazing, and may be sown from late September to early December. Southern grown seed must be used. If Northern seed are sown the leaves will make a flat growth on the ground during the early winter and afford little grazing. The Southern grown seed will stand erect and give good grazing throughout the winter. During an average winter, rye will not give as much grazing, on most soils, as will oats. If the winter is quite cold, rye will frequently give a better growth than oats. About one and a half bushels per acre should be sown, either broadcast or in drills. Some prefer broadcast sowing with all grains designed for winter pasture. No definite experiments have been made along this line in Louisiana. Oats, rye and barley sown with a drill seem to withstand cold weather better than that sown broadcast, but broadcast sowing seems to give a more complete covering of soil in early winter.

Rye matures about a week to ten days in advance of oats, other conditions being the same. On the Experiment Station grounds we prefer oats to rye for winter pasture, but there may be some conditions under which rye would be preferred.

If one cannot secure home grown seed, Georgia seed are to be preferred. It is very important that Southern grown seed should be secured.

BARLEY.

Barley thrives in Louisiana, when the season is not especially favorable for the development of the grain rust. When the fall and winter season is dry and cool, barley may do exceedingly well, and from a single year's experience, lead one to think it the best winter grazing grass to be had. The following year it may be almost a complete failure. It is more variable in its habits, with us, than any of the grains that may be sown in the fall of the year.

Those wishing to try it should sow from one and a half to two bushels per acre, as they would oats, in early October, or even in late September. As with oats, planting may be made as late as early December, but the later the planting is made, the longer will be the period required to produce sufficient growth for good grazing.

If grazing is discontinued in early February a fairly good crop of grain may be secured, if the plants are not badly damaged by rust. Probably twenty bushels of grain would be a maximum crop, and ten bushels an average crop on the thinner soils, and better returns from the better soils.

WHEATS.

There is no justification in planting wheat for the grazing of hogs in Louisiana, as none of the varieties makes better grazing than oats, and all are more or less subject to the grain rust, and some years the wheat will be almost entirely destroyed in the late winter by a heavy attack of rust. It is possible that we may obtain a wheat that will make good winter grazing in the northern portion of the State, and then make a fair yield of grain, but for the present it is the part of wisdom to plant sparingly of wheat. For those who wish to try wheat, give preference to varieties grown in Georgia and Southern Tennessee. The Mediterranean is one of the best.

VETCHES.

The hairy vetch and the Oregon, or smooth, vetch have been extensively advertised. We have grown vetches for ten or twelve years, and have not yet felt justified in urging their adoption throughout the State. In the judgment of the writer, the hairy vetch is to be preferred to the smooth. It is better to plant them in oats than to plant alone. When good seed are secured, and favorable conditions prevail, both varieties make a good growth, and afford good grazing. For those who wish to try vetch, plant about thirty pounds per acre with oats, or sixty pounds per acre, if planted alone. In a few instances known to the writer, vetches have been successfully planted in Bermuda sod, without any plowing being done, or any attempt made to cover the seed. Tests at the Experiment Stations have not been successful after this fashion. The seed should be covered in regular planting to about the same depth as oats and so can be planted at the same time as oats. It has been our experience that if the vetches are grazed too close to the ground the plants will die.

The following is a summary of the good points of the vetches: They grow during the winter, are not killed by cold in the southern portion of the State and are not apt to be severely injured anywhere in Louisiana. They gather nitrogen from the air. They make a very rich hay, and fairly good pasture if the pasture is not overstocked. Where the seeds ripen they will remain over summer without germinating, but will grow the following fall, even if covered to moderate depth by plowing.

The objectionable points are: High cost of seed, unreliability of seed, and the fact that both seed and hay cannot be made from the same crop. Louisiana does not furnish favorable conditions for seed crop for market or enlarged planting, but the crop may reseed the land on which it grew if pasturing is discontinued early enough.

We have in Louisiana two or more wild vetches that are of some value for grazing for hogs. Their growth should be encouraged, in pasture lands and wood lots.

ALFALFA.

Where alfalfa can be successfully grown, it is a desirable crop for grazing hogs, and makes excellent hog feed in the form of hay. In fact, hogs eat alfalfa hay probably better than any other that we can grow, and they thrive on it when a little grain feed is added. Alfalfa should not be grazed very much during the periods when grasses and weeds thrive—that is, during the late spring and during the summer rains. The grazing animals eat the alfalfa only and the grass and weeds are given an opportunity to develop to an extent that enables them to soon overshadow the alfalfa, and eventually to kill it out. During good weather in winter the crop is not seriously injured by pasturing hogs on it, and the animals thrive with little or no other food.

Alfalfa can be grown on most of the alluvial lands along the Red River, the well-drained stiff lands of the Mississippi, and on many of the bayous throughout the State. Permanent water should not rise within three feet of the surface of the soil for any considerable period of time, else the alfalfa will be killed. In the southern portion of the State, fall planting is most desirable. In the northern portion of the State, fall planting is most desirable if favorable season can be secured, but dry weather is apt to interfere. Spring planting is therefore frequently resorted to with satisfactory results. It is best to use about thirty pounds of seed per acre, covering not more than an inch deep. Special effort should be made to have good drainage. On soil that is not naturally suited to alfalfa, a light dressing of stable manure, that has been sprinkled with soil from a successful alfalfa field, is recommended, though this will not always insure success. Grazing the early spring crop of alfalfa with hogs sometimes is resorted to for the purpose of delaying the first cutting, as early fall sowing is apt to give a crop ready for harvest before good hay making weather prevails. An acre of alfalfa will sustain from fifteen to thirty hogs weighing from one hundred and twenty-five to a hundred and seventy-five pounds, but it is probably best to not carry so many, if the alfalfa is to be retained for a long period.

The best use to make of alfalfa in connection with hogs is to pasture it during good weather in the winter time, and make hay of it during the late spring and summer, and feed the hay to the hogs at such periods as may be necessary to supplement other feed, or to become the main feed of sustenance.

BERMUDA GRASS AND WHITE CLOVER.

No better early spring and summer pasture can be secured at nominal cost than in Bermuda grass and white clover. Hogs do well on it. Bermuda grass sod is easily secured anywhere in the State. In many places it is sufficiently well established to make a sod in one year if allowed to do so. Where it is not established, the field may be planted in corn and after the last cultivation small tufts of sod planted at intervals of a few feet in the corn rows, and Bermuda will be established pretty well by the end of the season, and the second year will make a complete sod early in the summer. White clover seed can be sown at the rate of four or five pounds per acre in this sod in the early fall, and sometimes a set may be secured the first year. If not, the clover will reseed the land, if not pastured excessively during early spring, and thereafter the white clover will make very early spring pasture that will give good service in the way of hog feed, and will last until late spring, when the Bermuda becomes good. This combination comes nearest to being perpetual pasture of anything that we have been able to find, and we have tried all combinations that seemed to give any promise of meeting this desire. We have found that crops which thrive during the winter and make a heavy growth, crowd out the Bermuda and leave the land bare in large spots in early summer. Spotted Medic, and other burr clovers, have not been as satisfactory for combination purposes, though they make a more vigorous growth in early winter. Animals have to be trained to eat the burr clovers.

SUMMER CROPS FOR HOGS.

SORGHUM.

The sweet sorghums make a very acceptable forage for hogs in the early summer, coming at a time when it is sometimes difficult to have other green feed. Sorghums are most serviceable for grazing during May, June and early July. The results of experiments at the station have not indicated the high value of sorghums as a grazing or soiling crop for hogs that they are reputed to have. In fact, we do not place a very high value on them, and commend their use only as a makeshift. The only strong points in their favor are that they give quick returns, and may be had about the time winter crops are exhausted and summer crops are not sufficiently matured for best service. The Early Amber, Early Orange or Coleman are to be preferred. Any of these may be sown in drills as soon as danger of frost is passed, though growth will not be rapid until warm weather prevails. The middle of March to the first of April is ordinarily a good time to plant sorghum. However, the crop may be planted as late as the first of August. When sown in drills, with rows three and a half feet apart, it takes about twelve pounds per acre to give a thick stand. It is best to plant the seed thick so the stalks will be small. The crop should be well cultivated until it is about two feet high, if it is to be grazed, or as conditions may require, if it is to be soiled. Planted in early spring, the sorghums will mature in about a hundred and ten to a hundred and twenty days. Planted in late June or early July, they will mature in sixty-five to seventy-five days. Soiling may be begun as soon as the stalks begin to head, but are more valuable as the seeds mature. Soiling involves considerable labor, but is sometimes to be preferred. In many instances the batture along the Mississippi River offers considerable areas of rich soil that is useless except for a crop of quick maturity. Sorghum may frequently be planted to advantage on such land, and grazed or soiled to hogs. Sometimes sorghums may be worked in as catch crops, between other crops, because of their early maturity.

SWEET POTATOES.

Sweet Potatoes grow well in all portions of the State, and in all soils except occasional areas that are excessively rich in nitrogeneous matter. Sandy soils in the alluvial lands are generally better suited to potatoes than stiff lands. In the hill lands, the loams and mixed sandy soils are the best. In some localities, distinctly clay soils produce the best potatoes. This crop re-

EXPERIMENT STATION HOGS ON SWEET POTATOES.

quires only a moderate amount of moisture. Plants may be put out from early spring until the early part of August. When planted early in the spring, they do not produce large roots much in advance of plantings made later. For this reason they are especially adapted for planting after spring crops have been harvested. Oats and clovers harvested in May and June may be followed by sweet potatoes to advantage. If early plantings are made in small areas, they will afford vines for planting in June for extensive areas. The large yams or some of the large white varieties are to be preferred. The Southern Queen produces large yields. The yams seem to be preferred by hogs, though this has not been thoroughly established. A bushel of seed potatoes put in a hotbed early in the spring will give slips enough to plant sufficient area to afford vines for two acres of planting soon after oat harvest, or two or three times this area for late planting. Sections of vines containing three leaves make a good size for setting. Vines may be rapidly set when the soil is quite wet by pushing one end of the vine into the soil with a stick and packing the dirt around it, or in dry weather holes may be necessary to receive a little water just in advance of the setting. Sometimes a shallow furrow is opened with a plow and the vines dropped so they lie across the furrow with one end at the bottom and another furrow is thrown on the vines and a roller is run over them to compact the soil.

Planted in June and early July, they will be ready for feeding about the middle of October. Hogs turned on them then will root more potatoes than they will eat, but there will not be very much loss from this cause. An acre of potatoes should feed eight to ten hogs, one-year-old, for sixty days, if one uses some supplemental feed in the form of rice polish or rice bran. We have uniformly secured good results from feeding hogs in this way.

The writer considers sweet potatoes as pre-eminently the best root crop for hogs for fall and early winter grazing. The cut-over pine hill lands will likely have their agricultural development as a hog-raising country from the fact that these soils are pre-eminently suited to the production of sweet potatoes, peanuts and cowpeas, and they produce fair winter oats for winter grazing. That this combination of crops will produce the most profitable hogs that can be raised in the United States, is the firm conviction of the writer.

PEANUTS.

Peanuts will grow in any soil in Louisiana, but are most productive in the moderately sandy soils. The largest yields are secured from the grey sandy hill lands, and the red clays of the hills of North Louisiana, that are only moderately stiff. Peanuts are most excellent hog feed. Planted in early spring, they are

SPANISH PEANUTS.

ready for grazing about the first of August. Planted after oats and clovers in June, they are ready for grazing about the latter part of September. They come in very timely for feed between cowpeas and sweet potatoes. It pays to plant plenty of seed and secure a thick stand. About one bushel of shelled peanuts per acre is required of the Spanish or Virginia. The Spanish being smaller, will give a thicker stand for the same planting, but they will permit thicker planting than will the large types. Plant in rows as narrow as possible to permit comfortable use of implements in cultivation. Thirty to thirty-six inches is a good width. Hills four inches apart are not took thick, though eight to twelve inches is commonly recommended. More than twelve inches is too thin.

The Spanish has a decided advantage of the spreading varities in its erect habit of growth, permitting of easier cultivation, and when it is desired to harvest the tops for hay, they may be cut more readily with a machine before the peanuts are harvested or grazed.

It is desirable to feed hogs a small amount of corn when they are on peanuts to more nearly balance the ration.

An acre of peanuts should feed eight to ten hogs for thirty days or more, when the hogs weigh in the neighborhood of two hundred pounds.

If more peanuts are planted than the hogs can consume in the time desired to feed them, they may be harvested and afterward fed as dry feed to hogs or any other kind of livestock with good results.

The present price of peanuts is such as to render their culture for market very profitable in soils best suited to growing a good nut that can be harvested clear of adhering soil, and where this crop is raised as a money crop, only enough hogs should be maintained to consume the inferior pods and clean the fields of what would otherwise be wasted. It is very probable that the peanut crop will increase until the market is somewhat depressed and then the crop will be profitable when converted into pork. Early planting may be made for early grazing. Hogs fed on peanuts produce meat that cures with a darker color than that obtained from corn-fed animals, but hams from hogs fed on peanuts and finished on corn for ten days or so before slaughter, are not to be excelled anywhere.

Peanuts greatly enrich the soil in nitrogen, and when the crop is pastured the land receives from the crop a very appreciable amount of vegetable matter. No more desirable crop can be grown in many portions of Louisiana, where conditions are peculiarly favorable to the growth of this most excellent feed crop.

CASSAVA.

Cassava is well suited to the sandy lands of the State. Short sections of the stems are cut before the period of danger of frost, and placed in moist sawdust, as the horticulturist treats cuttings from grape vines, and these sections are planted in the early spring. The crop occupies the land all summer and requires cultivation until late summer. Plant in rows four feet apart, and have the hills about two feet apart in the row. The root only is valuable for feed. These roots cannot be harvested more than a week or ten days in advance of feeding, as they will rot in a short time after being removed from the soil. Very satisfactory gains have been reported in Florida from feeding cassava to hogs and to cattle. In Louisiana we have found that we can make a larger tonnage of sweet potatoes than we can of cassava, and make the crop at about one-third the cost. We do not consider that cassava should find a place on the average farm in Louisiana, but we receive so many inquiries about the plant that this data is given here. There are also those who wish to try it for their own satisfaction, and these brief suggestions for cultivation are given. The cuttings may be secured from most of the nurseries of Florida, names of which can be secured from the Experiment Station.

SOY BEANS.

The rapidly increasing popularity of soy or soja beans in Virginia, Tennessee and some other sections of the country has caused a great deal of inquiry about soy beans for Louisiana. We have grown a large number of varieties for a number of years, and have not yet secured results that lead us to be very enthusiastic about them. It seems very probable, however, that some of the hill lands may be devoted to soy bean culture with moderate profit, either for grazing hogs, or for the seed market, allowing hogs to follow the harvest and gather the shattered beans. It will be well for the farmers throughout the State to try the Mammoth Yellow, the Ito San, and possibly one or two other varieties. They may be planted from early spring to late summer. Probably the best use to make of them is to plant them after oats or other spring crop, planting in rows no wider than will permit of cultivation, and cultivate two or three times. They thrive well on rather poor soils. On rich soils, the yield of fruit is less but the growth of vine more rank. It will require about one-third to one-half bushel per acre for planting.

When seed are desired, the crop should be cut as the beans begin to ripen, and before the pods begin to split open. They may be thrown in small cocks, after wilting, and allowed to remain two or three days, and then stacked in an open shed or according to special methods. Hogs may then be turned into the field to gather the shattered beans. Soy beans ground and mixed with corn make a most excellent feed for hogs, as reported by some feeders. If they are to be used as forage for hogs, they should be pastured as soon as the pods are fully matured, and before the seeds shatter on the ground.

For hay, soy beans will be more desirable under some conditions than cowpeas, but for forage for hogs the writer prefers the cowpeas.

COWPEAS.

Cowpeas are well known throughout Louisiana. They are most commonly planted in corn at the last plowing for fertilizing the soil, and for hay. They should be almost universally planted in corn. If planted alone, early spring planting (late March and early April) may be made broadcast or in drills; if in drills, they should be cultivated once or twice. Broadcasting alone or in corn, at last cultivation, requires from one to two bushels of seed per acre. In drills, three pecks to one bushel is ample.

Cowpeas make very good pasturage for hogs. They are ready for pasturage at a time when we need them. When planted early, they may be brought in before peanuts, or the cowpeas in corn may be pastured for a month before peanuts, following oats, are ready.

Hogs may be allowed to gather both the corn and the pea crop profitably, if portable fences are used, so that the animals

COWPEAS ON EXPERIMENT STATION.

can be confined to an area that they can harvest in about ten days.

When first turned into corn and peas, hogs will break down more corn than they will consume, but later the shattered corn will be eaten by them, so that the losses are small. Corn may be harvested in this way before it is sufficiently dry to be gathered for the barn. Hogs may be put on the average crop about the last week in July, and kept there until the first of October. In that portion of the State where a heavy crop of seed is produced, there may be danger of hogs consuming too large a quantity of the seeds.

The New Era pea is becoming popular where the Clay and Speckled peas do not fruit well. This variety generally makes a pretty good crop of seed and a fairly vigorous vine, and it matures early.

Cowpeas are an exceedingly desirable crop as a soil renovator, a hay crop, and as a grazing crop for hogs.

GROUND ARTICHOKES, OR JERUSALEM ARTICHOKES

The ground artichoke thrives in rich soils in all portions of Louisiana. A sandy loam soil with plenty of humus gives the best crop. If grown for hog feed, they should be planted in rows about four feet apart, and the hills about a foot to a foot and a half apart in the row. They must be planted as soon in the spring as danger of frost is past. It requires from three to eight bushels to plant an acre, according to the size of the tubers planted. Plant as Irish potatoes are planted. It is not always easy to get a stand, but they are vigorous growers when once started.

The seed are expensive, the crop requires considerable cultivation, and hogs are not as fond of them as some seed catalogues and advocates of their planting would have us believe. Sweet potatoes meet every requirement that the artichoke would meet, make as large returns, occupy the land a shorter time, and are less difficult to handle in every way, as well as being a less expensive crop in the matter of seed and the disposition of the tops. We, therefore, see no reason for advocating the cultivation of artichokes for hogs, but have given the above data because we have frequent letters inquiring about the crop.

SUPPLEMENTAL ROOT CROPS

Winter-growing root crops may be used to advantage as a supplemental feed throughout the greater portion of the State. Rutabagas and stock beets are the best crops for this purpose, although stock carrots may also be profitable, especially in the northern portion of the State, as they are not as susceptible to freeze as are other root crops.

STOCK BEETS.

On rich land stock beets will make as large tonnage of succulent feed as will any crop that can be planted for winter growth. In the southern portion of the State, on good land, twenty-five to forty tons of beets may be secured, if they are planted in early October and allowed to grow until the following April or early May. It is best, however, to plant them in early fall and begin feeding as soon as they are large enough, which will ordinarily be about one hundred to one hundred and fifteen days after planting. If the planting is made the latter part of September, or early October, we can begin feeding them about the middle of January and continue until midsummer, if desired, gathering from the field enough at one time to feed for several days. The roots will keep a week or ten days after they are harvested. Hogs may be maintained on beets alone, but it is advisable to use them only as a supplemental feed, and in this way they can be used to advantage. When bad weather is anticipated, one can harvest a quantity of beets and feed them to the hogs during times that pasturing the fields would be injurious to the land, as well as exposing the hogs to inclement weather.

Plant about eight pounds per acre. The seed at present are high, costing about twenty-five cents a pound. Plant in ridges, convenient distance apart for cultivation, and drill in the row as garden beets and later thin to a stand of one plant to every ten or twelve inches, or plant a number of seeds together in hills a foot apart in the row and afterwards thin to one plant in a hill. The seed do not have a high percentage of vitality, hence it is necessary to plant thicker than one desires the plants.

Late plantings may be killed by freezing temperature. A sharp freeze will kill the outer leaves of large beets, but the new

STOCK BEETS ON EXPERIMENT STATION.

TURNIPS.

leaves quickly come out and the plant recovers rapidly from the damage. Cultivate with any convenient implement to keep the soil mellow and free from weeds and grass. Begin feeding as soon as the roots are large enough, gathering as demands may require. A hog weighing a hundred and fifty pounds will eat from ten to fifteen pounds of beets per day.

In the southern portion of the State beets and rutabagas may be sown on early fall plant cane, and be harvested before time to offbar the cane in the early spring.

Giant Long Red is the largest yielding variety with us, although the other varieties do well.

RUTABAGAS.

Rutabagas may be fed to advantage, also, as a root crop for hogs. They develop more rapidly than do beets, but do not give such large yields. They may be planted earlier, and therefore may be used conveniently to precede beets as a source of succulent feed. In the northern portion of the State we can plant the seed in July and August, and have roots large enough to feed in about eighty days. We consider it best to sow them in rows and cultivate them well. They should only be planted in moist rich land. Land that will make forty bushels of corn should make fifteen tons of rutabagas. In the extreme northern portion of the State the crop may have to be housed for use during January and February. Three to four pounds of seed per acre should be planted, and the plants thinned to a stand.

STOCK CARROTS.

Stock carrots will withstand a lower temperature than will stock beets or rutabagas. For that reason they may be more desirable under some conditions. We have found that White Belgian and the large Yellow Victoria stock carrots will produce from twelve to fifteen tons of roots, in rich, mellow soil, in four months' time. Late September or early October is the best time to plant them, although they may be planted most any time up to January, although the late planting will give much smaller yields. Plant in ridges and leave the plants quite thick in the drill. Cultivate as may be required. Hogs are quite fond of carrots, after they have become accustomed to eating them, and good results have been secured in feeding them. They may be fed from February to the latter part of April.

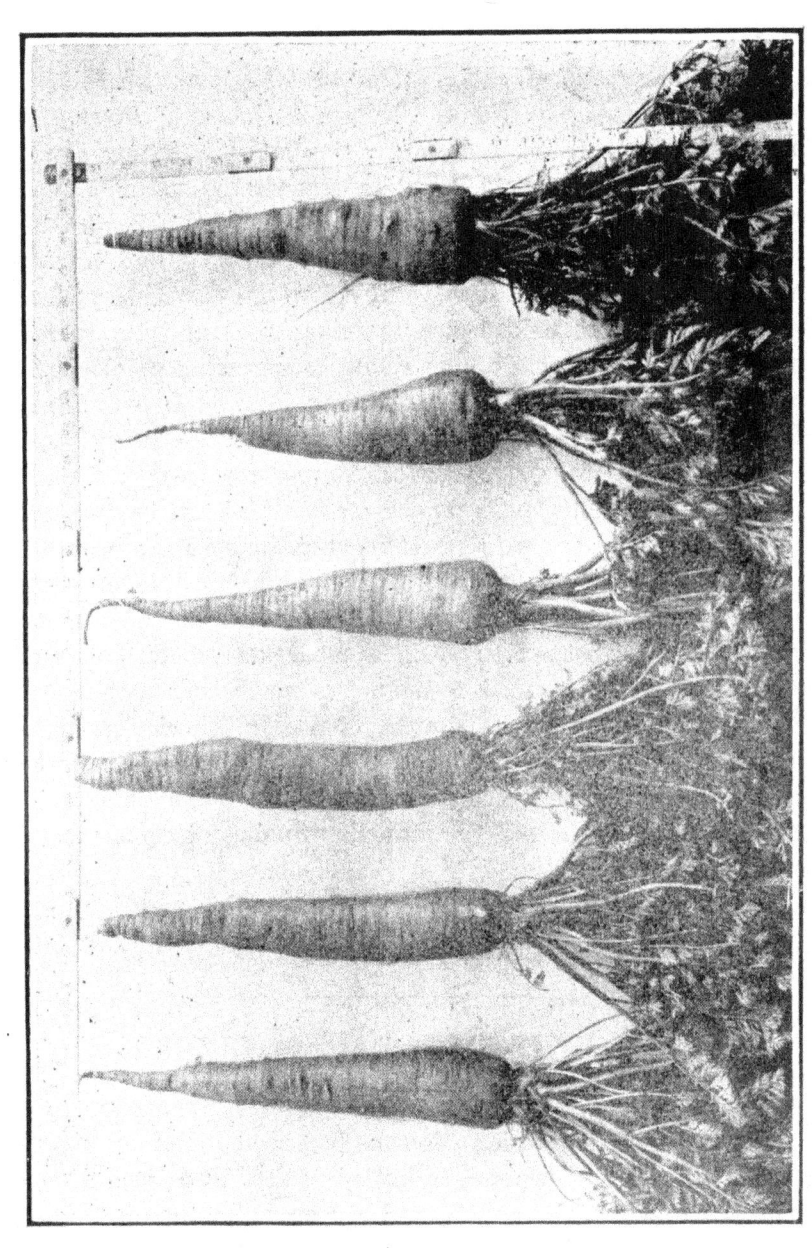

STOCK CARROTS.

ROTATION OF CROPS FOR HOGS

After many experiments we have come to the conclusion that the following succession of crops will give the cheapest pork in the greater portion of Louisiana: Oats and clover to be sown in separate fields in October, the oats grazed until early February, when the hogs should be transferred to the clovers, allowing the oats to make a crop of grain in May. Pasture the clovers until the middle of April or first of May, and transfer the hogs to Bermuda pasture, and allow the clover to make a crop of hay. Two fields sown to mixed oats and clover may sometimes be preferable to the single crops, using one field for early pasture and grain, the other for late pasture and hay. Keep hogs in Bermuda grass until cowpeas are ready to be grazed, in July for early peas or August for later plantings; pasture cowpeas until peanuts are ready, this crop having been planted after the oats are harvested. Continue on peanuts until sweet potatoes are ready for grazing, and continue on sweet potatoes until the hogs are ready for the market. When the hogs are on cowpeas, feed them a moderate amount of corn. When they are on peanuts feed half as much corn as when on cowpeas. When on sweet potatoes feed as much rice polish or rice bran as they were fed corn on cowpeas. The feeding period actually begins with the cowpea pasture, as the main purpose up to this time is to keep the animals growing nicely at as low cost as possible.

There will be times when a little grain ration may be necessary. One must exercise judgment in matters of this kind.

PLAN OF THREE-YEAR ROTATION.

The following succession of crops has been found by experiment to furnish a desirable continuity of feed and form a good series for easy rotation: Oats and clover, sweet potatoes, corn and peas, oats and clover, peanuts; then going back to oats and again following the same series. This gives the land a complete series of these crops in three years, and affords on three fields every crop each year. To carry out the plan one should divide the area to be devoted to these crops in approximately three equal fields. In field one, sow oats and clover in October or early November; in field two, plant corn in early spring and sow cowpeas in the corn at the last cultivation; in field three, sow oats and clover in early fall and the succeeding summer plant peanuts. Each field will then have the same series of crops, but beginning at a different point in the cycle.

To recapitulate, the following may make this plan more explicit:

Crop 1—Oats and clover.
Crop 2—Sweet potatoes.
Crop 3—Corn and peas.
Crop 4—Oats and clover.
Crop 5—Peanuts.

The series in field one would be crops 1, 2, 3, 4, 5, 1, 2, etc.
The series in field two would be crops 3, 4, 5, 1, 2, 3, etc.
The series in field three would be crops 4, 5, 1, 2, 3, 4, etc.

PLAN OF FOUR-YEAR ROTATION.

A four-year rotation can readily be arranged as follows:

First year—Corn and peas, followed by fall sown oats, or mixed oats and clovers.

Second Year—Oats and clovers, fall planting of root crops.

Third year—Roots harvested, peanuts, fall sowing oats and clovers.

Fourth year—Oats and clovers, followed by sweet potatoes.

Fifth year—Return the same crops as the first year.

The same plan of rotation should be followed in four different fields, thus giving each set of crops every year—that is, Field No. 1 should start out with corn and peas, Field No. 2 with oats and clovers, Field No. 3 with root crops, followed by peanuts, and Field No. 4 should start with oats and clovers, to be followed by sweet potatoes. The succession is then easily followed, and if the same succession is followed in every field the crops will be given in regular succession, and no confusion can possibly follow. This is not at all a difficult matter if one will give it a few minutes' consideration and have the plan well understood.

The above plan is the one now being carried out for hog experiments at the Experiment Station at Baton Rouge, and the one that we consider the best under average conditions, when hog feed alone is considered.

If this plan is not perfectly plain, write the crops in order given on the rim of a circle, and it will then be easy to read around the circle, starting from any desired crop.

METHOD OF ESTIMATING AMOUNT OF GRAZING THAT CAN BE SECURED FROM A CROP.

We are so frequently asked to state how many animals can be pastured on a given crop that the following is given as the basis for making a guess, for, at best, only a guess can be made, as there are so many variable factors entering into the consideration.

Suppose a man has a cowpeafield that he wishes to graze, and wants to know how many hogs to put on it to graze it down in a given time, or if he has a given number of hogs he wishes to provide grazing for and wants to plant accordingly: Three to four weeks will ordinarily be the most favorable period for grazing cowpeas. Most men make a fair guess at how much hay a given area would make if harvested for that purpose: Let us suppose cowpeas would make two tons of hay per acre; there would, then, be the equivalent of about one ton of leaves, stems and pods that the hogs would consume, as a minimum, and a ton and a half as a probable maximum, under most favorable conditions. We then consult the standards of feeding and see how much a hog of a given weight will require. For instance: We find that a hog weighing 170 pounds, such as we would have at nine months old, under the pasture system of raising hogs, will require, per day, 4.6 pounds of digestible dry matter, of which .58 pound shall be protein and 3.4 carbohydrate. Then, looking at our table of compositions, we find that a hog would have to consume the equivalent of ten pounds of dried cowpea vines per day to supply the necessary amount of carbohydrate, although there would be a large waste of protein. On the other hand, two pounds of corn and the equivalent of five pounds of cowpea vines dry would almost exactly furnish what the hog would require for fattening, and he would consume this quantity.

CORN AND COWPEAS.

Since the good grazing period would not exceed thirty days, we must have one-thirtieth of the crop consumed each day. If we suppose, as above suggested, that the hogs will consume one-half the total dry matter, which would be one ton, we must have enough hogs to consume about seventy pounds of dry matter per day, which would require seven hogs, if no corn is fed, or fourteen hogs, if we feed corn as suggested. Of course, if only one-half this yield is estimated, we would correspondingly reduce the number of hogs. This is about what we find from practical experience can be counted on. Of course, there are many variable factors that may influence the amount of grazing: the weather may be such as to cause the peas to shed their leaves rapidly, or second growth may begin, damp weather may cause vines on the ground to mould and be rejected by the hogs, etc.

An acre of cowpeas grazed by hogs—say, ten in number—should give from three hundred to three hundred and seventy-five pounds of gain in pork in twenty to twenty-five days' time, if two pounds of corn is fed each day to each hog.

If estimates are carried out in detail, as above suggested, it will be seen that the corn of one acre, and the oats from an acre, will feed hogs for a balanced ration for grazing two acres of cowpeas, one acre of peanuts and an acre of sweet potatoes, given in the suggested form of rotation, and very little, if any, feed need be purchased. Circumstances may make it profitable, however, to sell the oats and purchase rice polish. From experiments carried on at the Experiment Stations, it would seem that one may count on producing about seven hundred and fifty pounds of pork per acre per year, under this system of cropping. Some farmers will do much better than this; some will not do so well. This estimate is based on results secured on good land.

PORTABLE FENCES.

By S. E. McClendon.

It is convenient if one can arrange fields so as to afford proper rotation without building special fences. Even then, however, it is frequently desirable to pasture small areas, limiting the range by temporary fences.

To pasture small areas with hogs it frequently becomes necessary to have portable fences. One that can be easily moved

LINE OF PORTABLE FENCE.

and quickly set up, is made by taking three pieces 2 by 4 inches, 4⅔ feet long, and nailing five boards, 1 by 4 inches, 14 feet long—a convenient length—to them as shown in Fig. 1. These panels are supported by legs bolted to the top of the panels, as shown at A. When the panels are set up, the legs are pulled out, so as to brace up the fence, as shown in Fig. 1.

Two men can load this fence on a wagon and move it where needed. Shorter or longer panels may be made, to suit the convenience of the builder.

A panel of fencing of this pattern will cost as follows:

42 feet of lumber, at $18.00 per M	$ .75
A carpenter, at $2.00 per day, should build 18 panels, or, per panel	.11
3 bolts and nails, about	.15
Total cost per panel of 14 feet	$1.01
100 feet of this fencing will cost about	$7.00

Another good strong fence is made in panels 12 feet long. Each is composed of five boards four inches wide and nailed to three pieces 1 by 4 inches, by 3⅔ feet long, as shown in Fig. 2. This makes a fence 4 feet high when set up.

These panels are set in triangular frames, which serve as legs. The frames are made by crossing two pieces of boards 3½ feet long, about 6 inches from the top and nailing. They are then braced across the bottom by nailing a piece two feet long, as

FENCE PANELS.

shown at A, Fig. 2. A two-inch notch is cut at B, and in the center of the brace below at C, which holds the panels in place when set up. The wider the spread of these legs at the bottom, the stronger the fence will be. This fence can be braced down if necessary by driving a one-inch board 18 inches long in the ground near the center of the panel, nailing the panel to it. The fence will be securely held in place.

In a panel of this fencing, including legs, there are 26½ feet of lumber. At $18.00 per M. feet, this would cost 48 cents. A carpenter, at $2.00 per day, can build 20 panels, 12 feet long, which will make the panels, with nails, cost about 60 cents each, or about $5.00 per 100 feet.

The above fences may be used for cattle as well as hogs. Where hogs alone are to be pastured, a fence 30 inches high will answer.

For a temporary partition fence for hogs, a plain woven wire 26 or 36 inches high is perhaps the best and cheapest. This will turn hogs as long as feed is plentiful. If cattle are to be pastured in the same field, one or two barbed wires may be used above the netting.

ARRANGEMENT OF POSTS

For convenience in moving the wire, and at the same time protecting the wire and post, the following method should be used: set two posts in each hole at the ends of the line; set deep enough to hold the wire when stretched tight, which will be not less than four feet deep. In setting the posts leave a space of one inch between the two, as shown in Fig. 3. When ready to stretch the wire, run the end between the posts and hold it in position by bolting two pieces of 2 by 4 on the end of the wire, with the wire between them. Then pass the wire between the posts at the other end of the line, stretch as light as desired, and bolt two other pieces on the wire close up behind the posts. Then remove the stretchers and the wire is held as tight as if nailed.

Light posts may be driven every 15 or 20 feet between these corner posts to hold the wire up. If hard posts are used the staples should not be driven up so they can not be easily pulled when one desires to move the fence. To move the wire the staples are pulled out of the post, and the 2 by 4 removed from the ends, when the wire can be rolled up and moved where desired. One hundred feet of this fencing should not cost more than $2.50 or $3.00.

When one plans beforehand for pasture crops, it may be possible to arrange the crops in fields where the length would greatly exceed the width, and in pasturing off the crop use portable fences across the narrow field, at very little cost or inconvenience.

BREEDING CRATES FOR HOGS.

The dimensions of the box (Fig. 1) are: length, 5' 6", width 2' and height 3'. The length of the short box, which may be made by moving the end board j into the slot k, is 3' 6". The corner posts are 2" x 4" scantling and the sides 1" x 4" strips; a a a are joists for nailing the floor to; b b extra boards to which the joists are nailed to stiffen the sides of the box; c c are boar supports which hold the boar's weight during service. The one on the left is stationary, while the one on the right is adjustable to the size of the sow and should fit up tight against her side; d is a piece used to adjust the right-hand support; e is a pin which holds the support in place; f is a strip to hold d in the groove or mortise; the g's (of which there are six) are pieces that hold the supports solid and are 13" in lentgh; h is a woooden screw to hold the front end of the adjustable support in place; i is a ⅞" rod which is placed behind the sow to keep her from backing out of the box; j is a movable end board which is used to adjust the box to different length sows. When long sows are to be bred the board is placed in the end of the box, as shown in the diagram, and when the short sows are bred the board is removed and placed in the slotted board k. L L are cleats which hold the bottom end of the board j in place; m is a platform used to raise a small boar high enough to serve a large sow.

The accompanying illustration, shown in Fig. 2, is of the improved type. Instead of the adjustment for long and short sows being handled from the front of the crate that end is made stationary. Put in lower sideboards 10" high through which

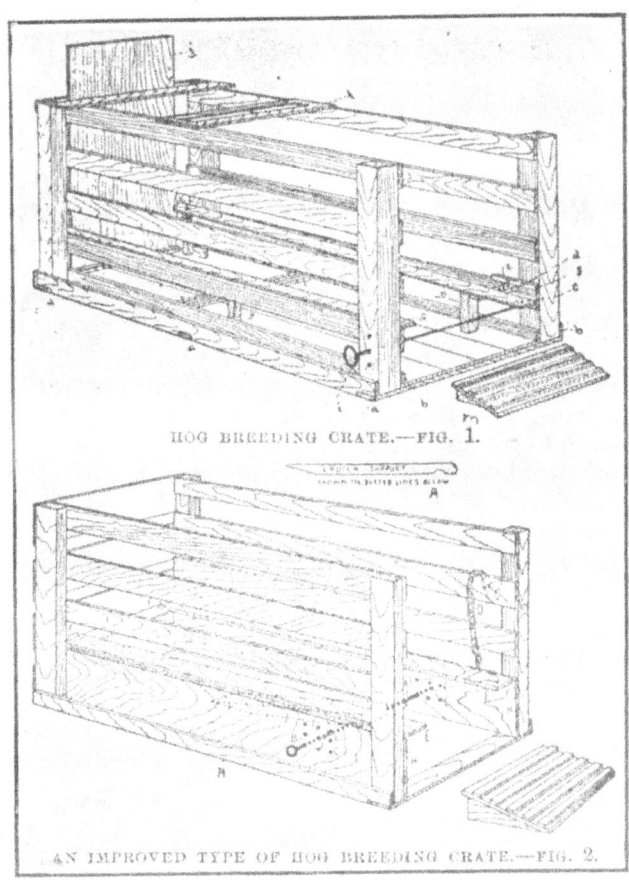

HOG BREEDING CRATE.—FIG. 1.

AN IMPROVED TYPE OF HOG BREEDING CRATE.—FIG. 2.

holes are bored at convenient intervals (C C C C) to admit the iron rod B, which should pass close under the hams of the sow just above the hocks. The proper hole to use is determined by the size of the sow. A crotch support A is added with a notch in it which passes between the sow's hind legs and rests on the retaining rod, as shown. This is 2" x 4" x 3' long, and the upper edges are rounded off smooth, so as not to injure the sow. The side supports for the boar E are made adjustable by hinging to one of the cross slats in front and are raised or lowered from the back by means of a chain (O) which passes over the top of side board, and fastens to a pin or heavy nail G. Put a chain on for each support. Two 4" boards, 6" apart, should be nailed over the top of the crate above where the sow's head comes to prevent her from climbing out.—*Courtesy of Breeders' Gazette.*

THE SCHULER METHODS OF CURING PORK ON THE FARM.

Col. Chas. Schuler, State Commissioner of Agriculture and Immigration, has, for years, employed the following methods of curing pork on his farm with such success that we take the liberty of reproducing them here:

When hogs are fat, select any time during the month of December, January or first half of February, when weather is clear, wind from the north to northwest, with the thermometer registering below 35 at sunrise. Have your water hot and scald as soon as hog is dead. Hang up and disembowel the animal just as soon as it is cleaned. No butchering animal should ever be permitted to cool off until after it is disemboweled. Cut up the carcass as soon as it is through dripping. Saw or split the backbone. Let it and the spare rib remain on the side, and make them as long as you can. Hams and shoulders small. Hams to sell readily should weigh from fifteen to eighteen pounds. Jowl will mix very nicely with trimmings and shoulder in making sausage, either for house use or the market. Feet, when cleaned properly, and put, raw, in strong brine, will keep all right for several months. Spread the joints and sides in your smoke house, applying a small quantity of salt to each piece. Let it lay until next morning to cool, then pack away, using plenty of clean salt.

To Sugar Cure Hams.—To half bushel fine salt add half pound pulverized saltpeter, one pound finely ground black pepper, four pounds brown sugar, mix thoroughly. Rub hams with mixture. Pack in box, skin side down. Apply double handful of mixture to flesh part of each ham. Then apply plenty of clean salt, never permitting the meat to touch, without salt being between. Covering all parts and filling every crevice, and let them remain in the salt six weeks.

How to Smoke Ham.—After being in salt six weeks, select a clear day, string each ham, and dip in a boiling solution of one pound borax dissolved in fifteen gallons of water and hang up high in a dark smoke house (the higher the better) and smoke, using green hickory wood. Smoke daily for two weeks or more, as preferred. By April 1 at latest, hams should again be dipped in boiling water, to cleanse them from all impurities, wrapped in paper, then cloth, and this painted with some cheap mineral paint. Hang up again and leave until used or sold.

To Make Good Sausage.—Grind your meat as fine as possible; don't have it too lean. Season with salt, ground black pepper, a good supply of pulverized soda crackers, not too much sage, and some red pepper tea. Well mix and stuff in sausage cases Cases can be secured from packing houses.

www.ingramcontent.com/pod-product-compliance
Lightning Source LLC
Chambersburg PA
CBHW060004230526
45472CB00008B/1935